Proven Strategies | Guaranteed Success

ACT

Crack the Code

Dr. Wycondia West

ACT

Crack the Code

WEST METHOD ACT

Copyright © 2017 by Dr. Wycondia West

ACT: Crack the Code

All rights reserved. This book is not to be stored in a retrieval system, reproduced, or distributed. The contents can be used for teaching purposes only. Request more copies of this book in writing.

Dr. Wycondia West
P.O. Box 716
Saraland, AL 36571

www.wycondiawest@yahoo.com
www.wycodiawest@gmail.com

* ACT® is a registered trademark of ACT, Inc., which was not involved in the production of this product and does not endorse this product.

WEST METHOD ACT

Foreword

Someone did it! Someone finally "cracked the code" to the ACT! In this book, Dr. West gives all students practical tips, strategies, and skills needed to score well on the ACT. The book is loaded with information that is easy for students at all levels and from all backgrounds to understand. If you are a high school student who has dreams of going to college, this book will start you on your journey to college admission.

Dr. West laid the groundwork for success on the ACT in this book. As a former high school teacher and current higher education administrator, I highly encourage high school students to purchase a copy of "ACT Crack the Code" today. It will be one of the best investments you will make toward your future.

Dr. Alecia L. Watt
Director of Educational Opportunity Programs
Greenville Technical College

About the Author

Dr. Wycondia West is a well-experienced educator who has been impacting the lives of students for over 19 years. Dr. West began her career as an elementary school teacher in 1998. She shifted gears when she entered the doctorate program at Walden University in 2007. This transition put her on a unique path to researching effective school programs for adolescents. In the summer of 2013, Dr. West decided to resign from the classroom in hopes of having a greater impact on all students. By leaving the classroom, she was able to provide a full-service tutoring program for K-12 students.

More importantly, Dr. West found her niche with ACT tutoring in 2013 when high school juniors and seniors began asking her for assistance. After successfully prepping a few students for the ACT, Dr. West decided to develop an amazing ACT test prep program that would benefit all students: high-performing and at-risk. The program became wildly successful and popular among individual clients. After having a consistent success rate, school systems started requesting Dr. West to teach ACT skills and strategies to groups of students. Now, Dr. West teaches her ACT skills and strategies to thousands of students in various school systems. Dr. West has cracked the code to the ACT!

From the Author

Every year, high school students prepare for the most important test of their high school career: the ACT. Sadly, most students are not clear on how to study or what to study. The ACT can be extremely intimidating; nevertheless, students can be highly successful if they have a guide and if they know which skills are covered.

This guide will give students a clear explanation of each section of the ACT: English, reading, science, math, and writing. Students will be crystal clear on skills, strategies, and time restraints. For example, students think they should take the ACT after taking pre-calculus or calculus; that's a myth. Pre-calculus and calculus skills are not tested on the ACT. Again, it is important for students to focus on specific skills, learn strategies, and increase their speed.

Each section of the ACT has specific strategies that students must learn to guarantee their success on this exam. Please understand that having a high GPA will not guarantee a high score on the ACT. This exam is 50% skill and 50% strategy. This guide will help students focus on tested skills and provide them with proven strategies that work.

ACT Up, Scholars!

Dr. Wycondia West
ACT Expert

Table of Contents

Testimonials……………..…………....……..…………1-6

Facts and Myths……………………..……..………....7-11

Section 1: English…..………………...…………….12-16

Punctuation Rules…………………..……..…………17-25

Important Conjunctions………………....……………26-28

Common Pronouns………………………...……………29-32

Section 2: Math……………..…..………..……….33-36

Math Rules………………….………...………..….37-41

Common Math Formulas…………....…....………..42-46

Exponents…………………….………..…………..47-49

Factorials………………………..……....…………50-51

Section 3: Reading………………..……………..52-56

Section 4: Science…..…...………..……………...57-60

Section 5: Writing……………..……………….61-64

Outline Worksheet…………………..…………….65-66

Sample Writing Prompts………..…...…..………..67-73

WEST METHOD ACT

Top Eight Tips...74-75

ACT Benchmarks......................….......76-77

Score High…...................…......…..........78-80

WEST METHOD ACT

Testimonials

WEST METHOD ACT

Testimonials

• Dr. West is wonderful! I highly recommend the ACT prep courses. My sons ACT score in English increased by 10 points. After working with Dr. Wycondia West, my son was awarded the Chancellor's Award at Troy University.

---Jennifer Luker

• I would highly recommend Dr. West ACT services. She is very knowledgeable and able to teach strategies that help students move quickly through the ACT. My daughter was able to go from a 23 to a 28 on her ACT after tutoring with Dr. West. With additional tutoring, she was able to get a 30 on her last ACT.

---Angela Palmire

• Dr. West is very passionate about what she does. Each student matters to her. Parents, if you're looking to increase your child's ACT score, look no further! Dr. West is the absolute best around! Because of her, my son's score increased by 4 points.

---Martha Jaquay

WEST METHOD ACT

Testimonials

●Well, Holly did not make a 30 on her ACT...she made a 31! She is so excited, and we are so proud! *(This student went from a 27 to a 31.)*

---Mrs. Tammy Stringfellow

●I never believed he could raise his score by 4 points. We are very excited. Thanks for whatever you did. I will recommend others to you. *(This student went from a 28 to a 32. Then this student took the ACT again and made a 35.)*

---Mr. Germany

●Bailey went up 2 points on her ACT! She made a 26 overall, but she made a 33 on English.

---Mrs. Sheffield

●We celebrated ACT scores tonight. The students who attended your sessions improved their scores. Some students improved by 5 points! Thank you so much!

---Dr. Byrd
Conecuh County Superintendent

Testimonials

- I highly recommend Dr. West for your student's ACT prep. Dr West is very passionate about improving ACT test scores. She doesn't only do group session, but she will do individual if need to ensure each student understand the strategy and techniques for taking the ACT. My daughter has a 32 on the ACT!

--Mrs. Monica White

- Dr. West was extremely nice and professional during my visits. She helped me improve my ACT score and feel more confident about the test. After going to multiple tutors, Dr. West helped the most. I would highly recommend her and her services.

--Jalexis Edwards

- During my tenure as a senior counselor at C. F. Vigor, Dr. West helped many students improve their ACT scores to qualify for academic and/or athletic scholarships. Dr. West is an invaluable resource to any student seeking to increase his/her ACT score.

---Mrs. Cheryl Robinson-Sutton
Mobile County Public School System

- Natasha made a 24 on her ACT. This was her first try on the test!

---High School Coordinator

WEST METHOD ACT

Testimonials

- Cailyn made a 30! Now, she's looking to apply for more scholarships.

---Mrs. Clemons

- I just have to shout out Dr. West and her ACT Virtual Bootcamp. Let me preface my comment by saying that for 4 years, I worked for The Princeton Review, the "premier" test prep company in the US. I'm here to tell you that Princeton Review has NOTHING on Dr. West. Her classes are full of so many effective tips and strategies. She delivers this knowledge in a way that keeps the student's attention from beginning to end, which is a pretty tough feat at 4 pm in the middle of the week. Her methods broke the big, intimidating ACT down into manageable chunks. Specific highlights for my son were her teaching style, the way she structured the course, and her pacing of the content. I highly recommend her ACT Prep classes to any student interested in ACTing up on the ACT!

--Mrs. Shayla Barnes-Holloway

- Dr. West did an excellent job preparing my son for the ACT. He took her group subject reviews and a one-on-one class and was able to increase his score from a 28 to a 34. I had heard great things about her program, but it far exceeded my expectations (and anything my older children had participated in). She cares about her students, has a great connection with them and is a great motivator. I highly recommend Dt. West to anyone looking for ACT prep.

--Mrs. Lauren Ramsay

Testimonials

•Dr. West has a true calling to help students achieve their goals. After attending her workshops, my son increased his composite ACT score 5 points!! Yes, you read that correctly. His scores went from a 27 to a 32!! With this accomplishment, my son will be able to go to the college of his dreams to pursue aeronautical engineering. We could not be more proud of him or more happy with Dr. West's services!

--Mrs. Wendy Pettis

•Dr. West is the BEST! My daughter was stuck on her math portion of the ACT. Try as she might, she could not get it over a 31. With Dr. West's instruction, she got a 35! This brought her " stacked ACT" to a 36 and her composite to a 35.

Dr. West doesn't just offer excellent tutoring, but she offers sound advice, encouragement, a cheer leader, a confidence booster, and support long after the exam is over. While we met professionally, I consider myself blessed to now count her among my friends. If you don't use her for your child's ACT needs, your child is missing out.

--Mrs. Tina Williams

Facts & Myths

WEST METHOD ACT

What people say	Fact	Myth	Answer
You will need a calculator to solve all math problems.		√	You won't need a calculator for all math problems, but don't leave yours at home.
You should always read the passages on the ACT reading section.	√		Yes, you should select 3 of the passages that you want to read. You should skim through your final passage.
You should always read the questions first on the ACT reading section.		√	Never read the questions first on the reading section. This is a big distraction and will cause you to waste time.
You should take the ACT without practicing just to get a feel for the test.		√	Never take the ACT without practicing first. This will be the biggest waste of money and time.

WEST METHOD ACT

What people say	Fact	Myth	Answer
You should take the ACT writing each time you take the ACT.		√	Most colleges want to see one writing sample. You don't have to keep taking the writing.
If you are running out of time, you should bubble something in.	√		Never leave anything blank. This is like giving points away.
If you have ADD or ADHD, you may qualify for extended time.	√		If you have ADD or ADHD, you should see your physician and school counselor to consider getting your time extended.
If your ACT score is too low, it will never increase.		√	Sometimes the lower your score, the more points you will gain.
You should work most of your practice tests online.		√	For most of your practicing time, you should print your practice tests or work in a book.

WEST METHOD ACT

What people say	Fact	Myth	Answer
You will score high on the ACT if you have a high GPA.		√	A high GPA does not guarantee a high ACT score. You will need to practice.
You should work with one ACT tutor before each exam.	√		It is not a great idea to prep with multiple tutors or companies. All tutors won't teach the same strategies.
You should take the ACT at least 4 or more times.	√		Taking the ACT 4 or more times is a great idea. This will give you the opportunity to possibly apply to a college that will Super score your ACT.
You should time yourself when practicing for the ACT.	√		After you are clear on the skills and strategies, you should time yourself on each practice test.
You can only take the ACT in high school.		√	Students can take the ACT as early as 7th grade.

WEST METHOD ACT

What people say	Fact	Myth	Answer
You must always take your exam at the same testing location.		√	You can take your test at any location. Sometimes it's a great idea to switch locations. Yes, you can also take your test in a different state.
You should carpool with friends on test day. This will get you motivated for the exam.		√	It is a terrible idea to carpool with friends on test day. You shouldn't even test at the same location as your friends. This can be a huge distraction.

Section 1: ACT English

ACT English Overview

- The first section of the ACT
- The largest section (75 items)
- 45 minutes to complete
- 5 passages to edit
- Grammar
- Punctuation
- Word choice
- Eliminate repetition
- Save the long questions for last

ACT English

This section assesses the following ability:

- To make the text more clear
- To improve clarity or word choice in the passage
- To eliminate redundancy (repetition)
- To follow the rules of grammar and punctuation (commas, colons, semicolons, dashes, hyphens, apostrophes, subject-verb agreement, dangling modifiers, misplaced modifiers, and conjunctions)

Test Features

1. There are 75 questions
2. ONLY 45 minutes to complete
3. There are 5 passages to edit
4. Each passage has a title and 15 items underlined
5. Very often the option of "NO CHANGE" will appear on A or F

Special Tips

1. Your goal is to get all grammar, punctuation, and word choice items correct. That's usually about 60 items. If you can focus on those first, then you are more likely to get the other 15 items correct.

2. Don't just read the underlined section! You should read the sentence that includes the underlined section!

WEST METHOD ACT

This is the only way you'll know if the grammar, punctuation, or word choice is correct.

3. SKIP and COME BACK to items that require you to rearrange, delete, or add sentences in a paragraph. These are not grammar questions, so don't waste a lot of time trying to figure these out first. Skipping these items and working them last will help to increase your time. **WORK THESE LAST!**

4. Don't be afraid to select "NO CHANGE". Some items are correct.

5. Multiple choice items that seem too "wordy" usually WILL NOT be the correct choice. **LESS is usually BEST!** * However, you should only use more words when asked for more details or if it makes the sentence clearer.

6. Keep each passage in the same verb tense throughout. Remember, you are editing five different passages. Each time you change passages, pay attention to the verb tense. If the passage starts in past tense, then you should select verbs that are in past tense. If another passage is written in present tense, then you should select present tense verbs.

WEST METHOD ACT

7. Make an educated guess whenever necessary! DO NOT LEAVE Anything Blank! When the test proctor gives the "five minutes" call, you should start moving quickly so you don't run out of time.

Punctuation Rules

WEST METHOD ACT

Apostrophe

• use with contractions	**Who's** that knocking at the door? *Incorrect:* Is it **you're** turn to take out the trash? *Correct:* Is it **your** turn to take out the trash?
• use to show ownership/possessive	We need to figure out **whose** phone is ringing in the library. *Incorrect:* My dog chases **it's** tail every morning. *Correct:* My dog chases **its** tail every morning.

•**Note:** Whenever you see a contraction, be sure to say both words. Remember, apostrophes are not used to make words plural.

WEST METHOD ACT

Comma

• use a comma to separate phrases *Remember, phrases can be deleted from a sentence, and there should still be a complete sentence without it.	Only the tourists were allowed inside the cathedral**, which angered the locals.** *Incorrect:* For many years the hikers had to walk across the unstable bridge. *Correct:* For many years**,** the hikers had to walk across the unstable bridge.
• use a comma to separate items in a list	Franklin forgot to **shut the door, turn off the light, and feed the cat**. *Incorrect:* The rain forest is filled with beautiful strange creatures *Correct:* The rain forest is filled with beautiful, strange creatures.

WEST METHOD ACT

Comma

●use a comma to join 2 complete sentences with a conjunction *(for, and, nor, but, or, yet, so)*	Yesterday's trip was awesome, so we went back for more. ***Incorrect:*** The house was dark and we were afraid ***Correct:*** The house was dark, and we were afraid.
●use a comma to avoid confusion	***Incorrect:*** For most the year is already finished. ***Correct:*** For most, the year is already finished.

•**Note:** Commas are always overused on the English section of the ACT. Be sure to learn the rules and practice.

WEST METHOD ACT

Semicolon

• use to join 2 complete sentences	Butterflies in the meadow make a great scene; they move so gracefully. ***Incorrect:*** My parents overslept; so, we missed our flight. ***Correct:*** My parents overslept; therefore, we missed our flight.

• **Note: Don't ever use the following words with a semicolon: for, and, nor, but, or, yet, so.** These are conjunctions that are used with a comma to connect two sentences. These conjunctions should never be used with a semicolon.

Semicolon Words

Therefore

Nevertheless

Consequently

However

In fact

WEST METHOD ACT

Colon

●use after a complete sentence THEN there should be a list or an explanation to follow	Cheerleaders have some of the same things**:** bows, pom-poms, and lipstick. ***Incorrect:*** Suddenly, dad had to**:** turn around because he left a few things behind his suitcase, license and cell phone. ***Correct:*** Suddenly, dad had to turn around because he left a few things behind**:** his suitcase, license, and cell phone.

WEST METHOD ACT

Dash

• use to separate parts of a sentence • use to insert additional information	The high school—old and small in the middle of town—was abandoned. ***Incorrect:*** Yes, I'll come to your house only if I can park in the garage. ***Correct:*** Yes, I'll come to your house—only if I can park in the garage.

Incorrect:
The high school, old and small in the middle of town—was abandoned.

(This is an example of unparallel punctuation.)

Correct:
The high school, old and small in the middle of town, was abandoned.

WEST METHOD ACT

Hyphen

●use to connect words like a compound word ●use a hyphenated word to describe a noun *(person, place, thing, or an idea)*	Local restaurants hire **full-time** and **part-time** chefs for the weekend rush. *Incorrect:* Aunt Sarah needed to get a high, priced attorney who could probably win her case. *Correct:* Aunt Sarah needed to get a **high-priced** attorney who could probably win her case.

WEST METHOD ACT

Period

•use to show where a declarative sentence ends	Francine remained optimistic throughout the ordeal. ***Incorrect:*** Running down the hill. ***Correct:*** The pack of wolves came running down the hill.

•**Note:** It is important to recognize a complete thought/complete sentence. A complete thought can take a period at the end, a semi-colon between two thoughts, a colon then a list, or a comma with a conjunction.

WEST METHOD ACT

Important Conjunctions

WEST METHOD ACT

FANBOYS (for, and, nor, but, or, yet, so)

FANBOYS (conjunctions)	Meaning
FOR	cause or because
AND	in addition to
NOR	choice, option, alternative
BUT **YET**	opposition, contrast
SO	result or effect

Incorrect: I am sleepy, I need to go to work.

Incorrect: I am sleepy, so I need to go to work.

Correct: I am sleepy, but I need to go to work.

•**Note:** Remember, you are not only being assessed on grammar and punctuation, but you are also being assessed on word choice. Word choice matters.

•**Note:** When joining two complete sentences together with a comma, remember to insert a coordinating conjunction.

WEST METHOD ACT

FANBOYS *(for, and, nor, but, or, yet, so)*

Conjunction	Meaning	Sample Sentence
FOR	Because	I need to find a new job, for I don't like my current career.
AND	In addition to	I have one dog and three birds.
NOR	And not	Neither mom nor dad went to the jazz concert.
BUT	However	I go to work on Sundays, but I don't go on Saturdays.
OR	Either	Does Michael have any brothers or sisters?
YET	But	I really wanted my son to study at the university, yet he left the school before the semester ended.
SO	Therefore	My daughter is very kind, so everybody likes her.

WEST METHOD ACT

Common Pronouns

Common Pronouns

Who	• use as a subject to refer to a person or people *Example:* **Mr. Simmons** is the person **who** presented the awards. Who opened the back door?
That	• use to refer to things, teams, or groups *Example:* **The Vikings** is the team **that** won the championship. Where are the **books that** mom bought?
Whom	• use whom when "who" becomes an object to a verb or preposition *Example:* I do not know **with whom** I talked to yesterday. (with is a preposition) This is not the **girl whom** they intended **to fire**.

WEST METHOD ACT

Which	• use in a phrase *Example:* *Green Eggs and Ham*, **which** was written many years ago, remains a popular book among young children.

WEST METHOD ACT

Whoever	•whoever is used as a subject to refer to a person or people •whoever can be used to replace *he, she, I,* or *they* *Example:* Whoever double parked, you are being towed.
Whomever	•whomever is used in the place of an object to a verb or preposition •whoever can be used to replace *him, her, me,* or *them* *Example:* Give the correspondence to whomever you see first.

WEST METHOD ACT

Section 2: ACT Math

WEST METHOD ACT

ACT Math Overview

- The second section of the ACT
- The longest timed section (60 minutes)
- 60 problems to complete
- Work the difficult problems last
- Freshen up your algebra, geometry, and trigonometry skills

ACT Math

This section assesses the following skills:

- Pre-algebra
- Elementary algebra
- Intermediate algebra
- Coordinate geometry
- Plane geometry
- Trigonometry (10 to 12 problems)

Test Features

1. There are 60 questions.
2. You have ONLY 60 minutes to complete.
3. There are 5 answer choices instead of 4.
4. The questions will appear in random order.

Special Tips

1. Work the easiest problems **FIRST**! This will require some skipping around.

2. If you have no idea how to work a problem, then SKIP IT or use the ANSWER choices to work it backwards (plug-in the answer choices).

3. Some problems can be worked without using the calculator, but you shouldn't leave your calculator at home.

WEST METHOD ACT

4. Always use the process of elimination strategy. Mark off answer choices that you think are not correct.

5. Make an educated guess whenever necessary! DO NOT LEAVE Anything Blank! If you skipped anything, be sure to go back before time is up.

6. Your job is to study the special math rules and formulas listed in the following sections. These formulas and rules will not be given to you on the test.

WEST METHOD ACT

Math Rules

WEST METHOD ACT

Triangles	• The inside of all triangles equal 180 degrees. $a + b + c = 180$
Straight Lines	• All straight lines equal 180 degrees.
Supplementary Angles	• Supplementary angles equal 180 degrees. $115 + 65 = 180$

• **Note:** The inside of ALL circles, squares, rectangles, trapezoids, parallelograms, and quadrilaterals equal 360 degrees.

WEST METHOD ACT

Right Angles **Complimentary Angles**	●All right angles equal 90 degrees.
Exterior Angles	●Exterior angles of all regular polygons equal 360 degrees.
Perimeter	●Add all sides of the shape.
Percent	●A percentage should be converted to a decimal. *65%= .65 8%= .08*
Prime Numbers	●These numbers are only divisible by 1 and itself. 1, 2, 3, 5, 7, 11, etc.
Mode	●The number that is repeated more often than any other.
Mean	●The "average" of all numbers listed. Add all numbers and divide by how many numbers there are.

WEST METHOD ACT

Median	● This is the "middle" value or "middle" number when numbers are listed in numerical order from least to greatest. If two numbers are in the middle, get the average of the two numbers.
Range	● The difference between the largest number and the smallest number. Just subtract the two numbers.
Adding and Subtracting Fractions	● The fractions must have the same denominator before adding or subtracting. When working word problems, sometimes the fractions can be converted into decimals. This makes them easier to compute.

WEST METHOD ACT

<u>Dividing Fractions</u>	• **Keep, Change, Flip!** Keep the first fraction, change division to multiplication, and flip the second fraction.
<u>Multiplying Fractions</u>	• Multiply the numerator with the numerator. Multiply the denominator with the denominator. Don't forget to simply.
<u>Functions</u>	• $f(x) = 2x^2 + 6x - 3$ $f(4) = 2(4)^2 + 6(4) - 3$ $f(4) = 32 + 24 - 3$ $f(4) = 53$
<u>Logarithms</u>	• $\log_2 8 = 3$ $2^3 = 2 \times 2 \times 2 = 8$

WEST METHOD ACT

Common Math Formulas

Area

Triangle

Area = $\frac{1}{2}$ (base * height)

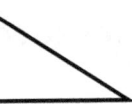

Square

Area = side ^{2}

Area = side * side

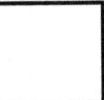

Rectangle

Area = length * width

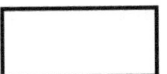

Trapezoid

Area = $\frac{1}{2}$ (base 1 + base 2)h

Parallelogram

Area = base * height

Circle

Area = πr^2

Circumference = $2\pi r$

Circumference = πd

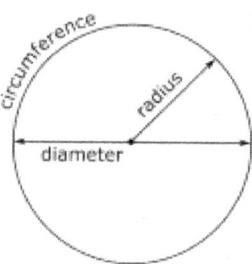

Rectangular Prism

Volume = length * width * height

Surface Area = 2(length * width) + 2(height * width) + 2(length * height)

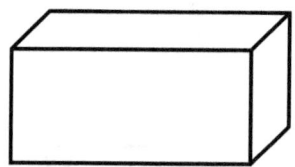

WEST METHOD ACT

Slope-Intercept Form	$y = mx + b$ where m = slope and b = y-intercept
Slope	$\text{Slope} = \dfrac{y_2 - y_1}{x_2 - x_1}$ where $x_2 \neq x_1$
Midpoint	$\text{Midpoint} = \left(\dfrac{x_1 + x_2}{2}, \dfrac{y_1 + y_2}{2}\right)$
Distance	$\text{Distance} = \sqrt{(x_2 - x_1)^2 + (y_2 - y_1)^2}$

Pythagorean Theorem

$a^2 + b^2 = c^2$

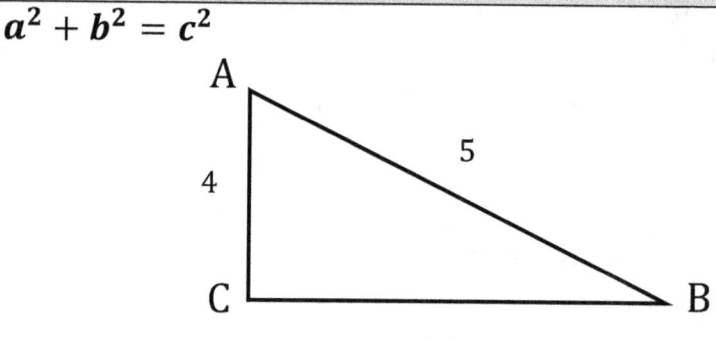

Trigonometry Formula

$\sin A = \dfrac{a}{c} = \dfrac{opposite}{hypotenuse} = \dfrac{3}{5}$

$\cos A = \dfrac{b}{c} = \dfrac{adjacent}{hypotenuse} = \dfrac{4}{5}$

$\tan A = \dfrac{a}{b} = \dfrac{opposite}{adjacent} = \dfrac{3}{4}$

Note: *SOH CAH TOA*

WEST METHOD ACT

Exponents

WEST METHOD ACT

Exponent Rules	Examples
When **MULTIPLYING** with exponents, **ADD** the exponents if the bases are the same.	$x^2 \times x^8 = x^{10}$ $2^2 \times 2^4 = 2^6 = 64$
When **DIVIDING**, **SUBTRACT** the exponents.	$\dfrac{x^5}{x^3} = x^2 \qquad \dfrac{x^2 y^5}{x y^3} = x y^2$
When **raising a power to another power**, **MULTIPLY** the exponents.	$(x^3)^2 = x^6$ $(3^2)^4 = 3^8 = 6,561$
EXPANDED power rule. When raising items in parenthesis to a power, everything is raised to the power.	$(x^2 y^4)^3 = x^6 y^{12}$

WEST METHOD ACT

When raising anything to the **ZERO POWER**, it is equal to +1.	$144^0 = 1$ $$(8^5 5^4)^0 = 1$$ Note: $(-12x^3y^5)^0 = 1$ $-(15x^4)^0 = -1$
When raising anything to the **NEGATIVE** power, write the **RECIPRICAL** of the number.	$5^{-2} = \dfrac{1}{5^2} = \dfrac{1}{25}$ $\dfrac{X^{-4}}{Y^{-7}} = \dfrac{Y^7}{X^4}$
More Exponent Examples ⟶	$5^3 \times 6$ $5 \times 5 \times 5 \times 6 = 750$
More Exponent Examples ⟶	$\dfrac{18x^5y^7}{22x^3y^8} = \dfrac{9x^2}{11y}$

WEST METHOD ACT

Factorials

WEST METHOD ACT

Factorials

n! is pronounced "n factorial"

If **n** is a positive integer, then **n!** is multiplied by all integers smaller than **n**

Examples

$0! = 1$

$1! = 1$

$2! = 2 \cdot 1 = 2$

$3! = 3 \cdot 2 \cdot 1 = 6$

$4! = 4 \cdot 3 \cdot 2 \cdot 1 = 24$

$5! = 5 \cdot 4 \cdot 3 \cdot 2 \cdot 1 = 120$

$6! = 6 \cdot 5 \cdot 4 \cdot 3 \cdot 2 \cdot 1 = 720$

Section 3:
ACT Reading

ACT Reading Overview

- The third section of the ACT
- One of the shortest sections (40 questions)
- 35 minutes to complete
- 4 passages to read

What Am I Reading?

1. **Passage one is always fiction.** This passage is typically from a short story or novel. You can expect to read about characters, encounter flashbacks, and some foreshadowing.
2. **Passage two is always social science.** This passage is typically historical, political, geographical, or deals with a social studies topic.
3. **Passage three is always humanities.** This passage can be characterized as a personal memoir or documentary. On the test, this passage typically has Part A and Part B.
4. **Passage four is always natural science.** You can expect to read about medicine, animals, space, or technology.

•**Note**: The passages will ALWAYS appear in this order. **You can read the passages in whichever order you desire**. When deciding on the order in which you will read the passages, think about the types of movies you like. This will help you decide.

ACT Reading

This section assesses the following ability:

- Reading and understanding of many styles of writing
- Referencing skills (questions that refer you to a few lines or paragraph)
- Inference skills/reasoning to draw conclusions

Test Features

1. There are 40 questions.
2. You have ONLY 35 minutes to complete.
3. There are 4 passages to read.
4. Each passage has 10 questions.
5. The 4 passages are identified at the top left corner in the test booklet: prose fiction, social science, humanities, and natural science.

Special Tips

1. Select the 3 passages that you will read first. Make this selection by reading the first 3 or 4 sentences at the beginning of each passage.

2. **DO NOT** read the questions first!

3. Mark Up Your Page. Be sure to underline or circle names, dates, locations, organizations, examples, and

WEST METHOD ACT

transition words in the passage. This will give your eyes something to focus on when searching for answers. **Do Not Annotate!**

4. Spend the first 30 minutes reading the 3 passages that you have selected. Remember, you can read the passages in any order. **Read quickly, and don't try to answer the questions without reading the passage.**

5. **The passage that you save for last, skim through it.** Whichever passage you think will be the most difficult for you, save it for last. This is called the **"5 Minute Passage"**.

6. If the question tells you to return to a certain line or paragraph, then RETURN to that line or paragraph.

7. Make an educated guess or use inferencing skills on some questions. DO NOT LEAVE Anything Blank!

8. Brush up on reading comprehension skills: author's voice/tone, main idea, theme, and vocabulary.

9. When answering **"Except"** questions, use Process Of Elimination to find which answer choice was **NOT** mentioned in the passage.

WEST METHOD ACT

10. Mark (+) or (-) on your passage to indicate positive or negative tones, attitudes, and examples. Marking your paper makes it easier to locate your answers.

11. **Main idea words...**

    - ✓ **Primary purpose**
    - ✓ **Primary focus**
    - ✓ **Main purpose**
    - ✓ **Main point**

Section 4: ACT Science

ACT Science Overview

- The fourth section of the ACT
- One of the shortest sections (40 questions)
- 35 minutes to complete
- 7 passages to complete
- Focus on the figures
- Focus on the tables
- Analyze figures and tables before answering any questions

•**Note:** **This is not a reading test!** The science section is not testing your ability to read these confusing passages.

WEST METHOD ACT

ACT Science

This section assesses the following ability:

- To think scientifically
- Interpret, analyze, and evaluate data *(figures and tables)*
- Draw upon some previous knowledge

Test Features

1. There are 40 questions.
2. You ONLY have 35 minutes to complete.
3. There are 7 passages. Six of the passages will have figures, tables, and keys. One passage will have only text.
4. Each passage will have 5 or 6 multiple choice questions. The passage that is primarily text will have 7 questions.

Special Tips

1. Skip the passage that does not have any figures or tables! ALWAYS skip this passage and complete it last. This passage is sometimes the most difficult, and it has the most questions.

2. Do Not Read the text first! Before answering questions, *ALWAYS* look over the figures and tables. Most of your answers will come from the data on the tables and figures.

WEST METHOD ACT

3. Locate key words for each passage: experiment 1, experiment 2, Figure 1, Figure 2, Table 1, Table 2 and any italicized words. Pay close attention to the numbers on the figures and tables. How are the tables alike? How are they different? See if the numbers on the tables are increasing, decreasing, or remaining the same.

4. Most of your answers will come from the figures and tables. The questions will typically reference a certain table, experiment, trial, or figure. Your job is to pay attention to the data and analyze it.

5. Some questions will be hypothetical *(what if and suppose)*. Base your answer on the information found on the tables. Pay attention to the pattern(s) and this will help you find your answer.

6. When you are tackling the more complicated questions, pay close attention to the answer choices and use (POE).

7. When it's time to tackle the passage that's all text, focus on each scientist's theory or student's hypothesis. On the "all text" passage, the questions will compare/contrast the theory or hypothesis of each scientist or student.

Section 5:
ACT Writing

ACT Writing Overview

- The fifth section of the ACT (optional)
- 40 minutes to complete
- This section will not increase nor decrease your composite score
- No penalties for your opinion
- Organize, organize, organize
- 4 by 4 writing strategy (write at least 4 paragraphs with 4 sentences)

•**<u>Note:</u>** Please start with an outline! A simple outline will help you focus on the prompt, develop your stance, and organize a great essay.

WEST METHOD ACT

ACT Writing

This section assesses the following ability:

- To organize your thoughts
- To provide your opinion on an issue
- To provide details to support your opinion
- To write in a logical manner
- To properly use grammar and punctuation

Test Features

1. A writing prompt is provided.
2. You ONLY have 40 minutes to plan and complete.
3. The prompt will usually ask for an opinion.

Special Tips

1. Always, always take the first 5 minutes to create a brief outline. You need these first few minutes to organize your thoughts.

2. Make sure your writing is legible. PRINT your essay if your cursive writing is not that great! Two readers will review your essay, so make sure they can read the essay.

3. The outline should have the following components: 3 points for your argument, supporting evidence, an opposing argument, and a supporting argument.

WEST METHOD ACT

4. **Paragraph 1**. After completing the outline, write your introduction on the writing paper. The introduction is simply the PROMPT restated. This is also your opportunity to include your perspective on the topic.

5. **Paragraph 2**. Point 1 on the outline paper should be written as Paragraph 2 on your essay paper. Provide a topic sentence to state your perspective. Now is the time to provide details to support your perspective. Discuss the perspective that supports your perspective. Provide details to show the connection and include personal evidence.

6. **Paragraph 3**. Point 2 on the outline paper should be written as Paragraph 3 on your essay paper. Select an opposing perspective. Discuss this opposing perspective and provide details to show why you oppose it. Personal evidence is a great touch.

7. **Paragraph 4**. Point 3 on the outline paper should be written as Paragraph 4 on your essay paper. Recap your discussion. Restate your perspective and arguments. Provide a final overarching thought on the topic.

•**Note:** Don't forget to add personal details to make your essay more interesting. Your job is to connect with the readers. Remember, two readers will score your essay. The highest possible score is a 12.

WEST METHOD ACT

Outline Worksheet

WEST METHOD ACT

Outline Worksheet

Introduction (restate the prompt + your perspective on the issue) _____

Point 1 Topic Sentence to state your perspective
_____

Supporting Evidence _____

Supporting Evidence _____

Supporting Evidence _____

Personal Evidence _____

Point 2 Topic Sentence to state the opposing perspective
_____

Supporting Evidence _____

Supporting Evidence _____

Supporting Evidence _____

Personal Evidence _____

Conclusion (Recap your discussion. Restate your perspective and arguments. Provide a final overarching thought on the topic.) You can provide five paragraphs if you are able to write them without getting off topic.

WEST METHOD ACT

Writing Prompts

WEST METHOD ACT

Prompt 1: High School Curricula

Currently, American high schools are emphasizing rigor and relevance across the curriculum. High school students are required to take and pass a specific number of core classes. High school students who are involved with various extra-curricular activities are also held accountable for passing extremely rigorous core classes. Today, many students find it difficult to maintain adequate grades with such a busy extra-curricular schedule. Is it important for high schools to hold all students to the same standard for core classes, or should they provide extra assistance for those students who are responsible for various extra-curricular activities?

Read and carefully consider these perspectives. Each suggests a certain way of thinking about high school curricula.

Perspective One	The rigorous and relevant core classes are essential to a quality education because they teach students how to think critically about a broad range of topics, thus preparing them to tackle college courses. Students who are involved in extra-curricular activities need to choose between excelling in their core classes or excelling in extra-curricular activities.
Perspective Two	Students who are involved in extra-curricular activities should be provided with additional support, so they can be successful in core classes. These students are typically athletes, cheerleaders, and band students. If these students fail their core classes, they will not qualify for scholarships to college.

WEST METHOD ACT

| Perspective Three | High schools should closely integrate core classes into their extra-curricular programs. For example, athletes should receive college credit that will go toward their college career. This would have a double impact: high school credit and college credit. |

Essay Task

Write a unified, coherent essay in which you evaluate multiple perspectives on high school curricula. In your essay, be sure to:

* analyze and evaluate the perspectives given

* state and develop your own perspective on the issue

* explain the relationship between your perspective and those given

Your perspective may be in full agreement with any of the others, in partial agreement, or wholly different. Whatever the case, support your ideas with logical reasoning and detailed, persuasive examples.

Prompt 2: College Football

College football is extremely popular in the United States. Rival games usually air on television and can rake in lots of money. College football teams have a large fan-base that travel across the country to show support. Of course, these college coaches and institutions reap the monetary benefit of the sport. Some larger athletic programs bring in millions of dollars to their schools. Even though most star athletes are awarded full scholarships because of their outstanding athletic ability, they are not offered funds that can be taken to the bank. Nevertheless, the coaches receive huge salaries and bonuses for big wins. In some instances, some athletes return home to poverty during off season. Given all of this, should colleges provide football players with a small salary during football season?

Read and carefully consider these perspectives. Each suggests a certain way of thinking about the role of athletics at colleges.

Perspective One	Colleges should support football players with a small salary. These players not only generate millions of dollars for schools, but they also help to increase the school's popularity among potential students.
Perspective Two	Colleges provide star athletes with full scholarships because of their athletic ability and not academic talent. Consequently, colleges don't really owe the athletes anything else. College is for academics, so athletes are participating at free will.

WEST METHOD ACT

Perspective Three	College football helps to create the overall experience of college life. Nevertheless, it is important for colleges to maintain its focus on academia. The fact that football coaches receive extremely high salaries and bonuses boggles the mind.

Essay Task

Write a unified, coherent essay in which you evaluate multiple perspectives on college football. In your essay, be sure to:

* analyze and evaluate the perspectives given

* state and develop your own perspective on the issue

* explain the relationship between your perspective and those given

Your perspective may be in full agreement with any of the others, in partial agreement, or wholly different. Whatever the case, support your ideas with logical reasoning and detailed, persuasive examples.

Prompt 3: Extracurricular Activities and Codes of Conduct

For many students, extracurricular activities are a meaningful part of the high school experience. These activities allow students to develop their skills in areas such as sports, music, and drama while building relationships with peers and gaining experience performing or competing. But at many schools, students who participate in extracurricular activities are subject to special codes of conduct. These codes often establish high standards for academic performance and behavior, and students must meet the standards to stay eligible for their activities. Should students who participate in extracurricular activities be subject to special codes of conduct?

Read and carefully consider these perspectives. Each suggests a certain way of thinking about the role of athletics at colleges.

Perspective One	All school rules and standards must apply equally to every student. It is unfair to hold student who play sports or music to higher standards than students who do not.
Perspective Two	Participation in school activities is a privilege, not a right. It is fair to ask students to earn this privilege by studying hard and behaving themselves.
Perspective Three	School programs should be open to all students. Not all students can meet high standards, which means not all students can participate in extracurricular activities.

WEST METHOD ACT

Essay Task

Write a unified, coherent essay in which you address the question of whether students who participate in extracurricular activities should be subject to special codes of conduct. In your essay be sure to:

* analyze and evaluate the perspectives given

* state and develop your own perspective on the issue

* explain the relationship between your perspective and those given

Your perspective may be in full agreement with any of the others, in partial agreement, or wholly different. Whatever the case, support your ideas with logical reasoning and detailed, persuasive examples.

WEST METHOD ACT

Top 8 Tips

WEST METHOD ACT

Top 8 Tips

1. Purchase a test prep book written by the makers of the ACT. Don't waste time trying to read through the book. Simply, use the practice tests.

2. Do not waste time reading directions for each section of the ACT.

3. Use a calculator that you are familiar with.

4. DON'T be a hero! You will not have to work out every single math problem.

5. Make sure you are timing yourself on practice tests.

6. On test day, the proctor (person facilitating the test) will give you a "5-minute call" before time is up. Work quickly, so you can beat the clock.

7. Practice reading material that's not very interesting, and time yourself.

8. Practice analyzing figures and tables on the science section (don't worry about answering the questions until you are reading to time yourself).

ACT Benchmarks

College Readiness Benchmarks

College readiness benchmarks were established in 2005 but updated in 2013. The benchmarks serve as success indicators for students preparing to enter college. In other words, if your subsection score falls below the specified benchmark, that's a red flag to colleges that you might have to enroll in remedial courses when you begin college. Many students begin college by taking remedial courses. Remedial courses are typically less rigorous and probably won't count as college credit.

Subsections	Benchmarks
English	18
Math	22
Reading	22
Science	23

WEST METHOD ACT

Score High

WEST METHOD ACT

5 Reasons to Score High

•College readiness indicator.
The national average on the ACT is a 21. The score on the exam reflects a student's college readiness. The Department of Education also determined that your ACT score is the second biggest indicator of your college ability. A higher score means you're ready for college.

•Receive merit-based aid.
Many colleges view a student with a high ACT score as worthy of additional financial aid or SCHOLARSHIPS. A high score indicates you're a good student. A high score can also get out of state fees waived.

•Avoid remedial classes.
Many colleges use your score to determine if you need remedial classes. Get a high score and save thousands of dollars in tuition.

•Improve your overall education.
Studying for the ACT can help you spot areas where you're weak in the classroom. Working on ACT test prep problems can sharpen your ability in school and improve your grades and impress college admissions officers.

WEST METHOD ACT

•*Perform well under pressure.*
Getting a high score on your ACT shows college admissions that you work well under pressure, and you're prepared for the challenging college curriculum. A high ACT score makes you a better college candidate. A high score can help you gain admission into the college of your choice.

www.ingramcontent.com/pod-product-compliance
Lightning Source LLC
Chambersburg PA
CBHW070808220526
45466CB00002B/600